Definitive Guide to *Cyber Threat Intelligence*

Using Knowledge about Adversaries to Win the War against Targeted Attacks

Jon Friedman

Mark Bouchard, CISSP

Foreword by John P. Watters

Contributions by
Jonathan Couch and Matt Hartley

Definitive Guide™ to Cyber Threat Intelligence

Published by:
CyberEdge Group, LLC
1997 Annapolis Exchange Parkway
Suite 300
Annapolis, MD 21401
(800) 327-8711
www.cyber-edge.com

For general information on CyberEdge Group research and marketing consulting services, or to create a custom *Definitive Guide* book for your organization, contact our sales department at 800-327-8711 or info@cyber-edge.com.

ISBN: 978-0-9961827-0-6 (paperback); ISBN: 978-0-9961827-1-3 (eBook)

Printed in the United States of America.

10 9 8 7 6 5 4 3 2 1

Publisher's Acknowledgements

CyberEdge Group thanks the following individuals for their respective contributions:

Editor: Susan Shuttleworth
Graphic Design: Debbi Stocco
Production Coordinator: Valerie Lowery
Special Help from iSIGHT Partners: Jonathan Couch, Matt Hartley, Patrick McBride

Table of Contents

Foreword

No professional sports team takes the field without scouting its opponent. No general launches a military exercise without studying the battlefield and the capabilities of the opposing forces. And no sensible business leader enters a market without identifying the major competitors and their strengths and weaknesses.

Yet every day most cybersecurity professionals go to work without any idea about the identity and probable actions of their adversaries.

In information security, if you do not understand the motivations, intentions and competencies of your opponents, then you cannot understand the risks to your enterprise or focus your defenses.

Eight years ago this insight led the founders of iSIGHT Partners to create the first firm dedicated to comprehensive cyber threat intelligence. We knew that enterprises could compete against increasingly focused and capable threat actors only by leveraging timely, accurate, relevant intelligence provided in actionable form.

We started the business to change the game in information security. Our aim has been to give cybersecurity professionals an alternative to constantly searching for vulnerabilities and trying to protect against all possible threats. We have pioneered new techniques for understanding the adversaries targeting our clients, and for exposing their tactics, techniques, and methods.

Today, cyber threat intelligence is a fundamental component of a world-class cybersecurity program. Yet the field is young, and few people understand it well.

That is why we are pleased to sponsor the *Definitive Guide to Cyber Threat Intelligence*. This short book provides an overview of the topic and its major activities: developing intelligence requirements; collecting, analyzing, and disseminating information; and using cyber threat intelligence to improve security at the tactical, operational, and strategic levels. It also offers insights into implementing a cyber threat intelligence program and selecting the right partner for that implementation.

I urge you to read this guide and share what you learn with your colleagues and your community. Cybersecurity has become more than a job. It is a discipline for protecting our enterprises, our livelihoods, the wellbeing of our customers and clients, and sometimes even our values and way of life. We cannot afford to let cybercriminals, hacktivists, and other threat actors stay a step ahead. Cyber threat intelligence is an essential tool for closing the gap and allowing us to thwart attacks before they cause damage.

I hope you will find cyber threat intelligence a stimulating topic, and this guide a useful resource for your own efforts in the field.

John P. Watters
Chairman and CEO
iSIGHT Partners

Introduction

"Cyber threat intelligence" sounds like a glamorous mashup of James Bond and Bill Gates, or perhaps Jason Bourne and Mark Zuckerberg. Indeed, today's headlines reveal a surprising number of master criminals and shadowy government agencies bent on world domination. Instead of stolen atomic devices, giant lasers, and exotic aircraft, they wield phishing campaigns, polymorphic malware, and DDoS attacks. The stakes are high, too: billions of dollars, personal information of tens of millions of customers and employees, and the protection of national infrastructures.

But information technology professionals don't have time for fiction (during working hours). They ask: "What, exactly, does 'intelligence' mean for cybersecurity? How can cyber threat intelligence help us do our jobs better? How can we design a cyber threat intelligence program?"

This book answers those questions. We describe the elements of cyber threat intelligence and discuss how it is collected, analyzed, and used by a variety of human and technology "consumers." We examine how intelligence can improve cybersecurity at tactical, operational, and strategic levels, and how it can help you stop attacks sooner, improve your defenses, and talk more productively about cybersecurity issues with executive management.

If you are concerned that the bad guys too often seem to be a step ahead of the rest of us, then please read on.

Chapters at a Glance

Chapter 1, "Defining Cyber Threat Intelligence," examines why cyber threat intelligence is needed, defines the term, and outlines its key characteristics and benefits.

Chapter 2, "Developing Cyber Threat Intelligence Requirements," explains the importance of developing good requirements related to assets, adversaries, intelligence consumers, and business operations.

Chapter 3, "Collecting Cyber Threat Information," describes the basic types of cyber threat information and how they are collected.

Chapter 4, "Analyzing and Disseminating Cyber Threat Intelligence," reviews the steps involved in preparing intelligence for different stakeholders, including security operations, incident response and executives.

Chapter 5, "Using Cyber Threat Intelligence," explores how intelligence can be used at the tactical, operational, and strategic levels to identify attacks and improve defenses.

Chapter 6, "Implementing an Intelligence Program," provides step-by-step recommendations for enhancing cyber threat intelligence capabilities.

Chapter 7, "Selecting the Right Cyber Threat Intelligence Partner," enumerates criteria for evaluating cyber threat intelligence providers.

The Glossary provides handy definitions of key terms (appearing in italics) used throughout this book.

Helpful Icons

Tips provide practical advice that you can apply in your own organization.

When you see this icon, take note as the related content contains key information that you won't want to forget.

Proceed with caution because if you don't it may prove costly to you and your organization.

Content associated with this icon is more technical in nature and is intended for IT practitioners.

Want to learn more? Follow the corresponding URL to discover additional content available on the Web.

Chapter 1

Defining Cyber Threat Intelligence

In this chapter

- Learn why targeted attacks are driving interest in cyber threat intelligence
- Define cyber threat intelligence
- Understand key characteristics and benefits of cyber threat intelligence

"There is nothing more necessary than good Intelligence to frustrate a designing enemy."

– George Washington

Cyber threat intelligence is a relatively young discipline. Most of the experts are employed by a handful of specialized cybersecurity firms, major government agencies, and large enterprises. Few people have a clear idea of the practices developed so far by the leading practitioners.

Yet the field is growing rapidly. Cyber threat intelligence is being covered by the press, and studied in depth by analysts at Gartner, Forrester Research, IDC, the SANS Institute, and the National Institute of Standards and Technology (NIST), among others. It is a subject of increasing interest to businesses and government agencies of all sizes.

This chapter discusses why cyber threat intelligence is a hot topic and provides a brief overview of its key elements.

The Need for Cyber Threat Intelligence

The surge of interest in cyber threat intelligence owes much to the devastating record of sophisticated targeted cyberattacks, including now-ubiquitous *advanced persistent threats (APTs)*. Even the largest, ostensibly best-protected enterprises have been victimized, sometimes to the tune of tens of millions of dollars.

The menace of targeted attacks

Ten years ago, IT security professionals mostly worried about *mass attacks*. Today these are regarded as secondary threats that merely generate "noise" on the network. For the most part, security vendors and enterprises defend against them successfully by analyzing the first instances discovered and quickly disseminating *signatures* and *indicators of compromise (IOCs)*. A few initial victims suffer, but everyone else can detect and block the attacks.

Today, the most serious data breaches and disruptions result from well-planned, complex attacks that target specific companies or industries. Sophisticated, well-funded attackers make detection difficult by:

- ☑ Utilizing social engineering techniques and multiphase campaigns that cannot be identified by simple threat indicators or blocked by frontline defenses.
- ☑ Constantly adapting their tools, tactics, and procedures to evade even advanced cybersecurity measures.

They have also raised the stakes by systematically targeting their victims' most valuable information assets and business systems.

The monitor-and-respond strategy

Most enterprises have recognized that signature-based defenses are not effective against sophisticated targeted attacks. They have shifted to a defensive strategy that focuses on monitoring and incident response.

The typical process can be summarized as follows:

1. Collect as many signatures, threat indicators, and security events as possible.
2. Feed this data to security products that can block mass attacks, such as next-generation firewalls, antimalware and endpoint protection software, and intrusion detection and prevention systems (IDS/IPS).
3. Use the same data to generate alerts, and monitor those alerts in the *security operations center (SOC)* with a *security information and event management (SIEM)* solution.
4. Have the SOC analysts examine the alerts, perform triage, and escalate the most serious to the *incident response (IR) team* for validation and analysis.
5. Have the IR team investigate the serious alerts, dig around in various logs, and reconstruct the elements of the complex attacks.
6. Use the attack analyses to stop the progress of ongoing attacks, clean up compromised systems, and protect against new instances.
7. Periodically report to the CISO the number and types of attacks so he or she can ask executive management for a bigger security budget.

Why the strategy is failing

Unfortunately, this process is rife with difficulties at all levels: tactical, operational, and strategic (see Table 1-1).

Tactical level

At the tactical level, unverified and trivial IOCs generate false positives and a flood of alerts; this noise prevents SOC analysts from identifying alerts linked to threats capable of causing real damage.

Operational level

At the operational level, it takes IR teams too long to find relevant information about threats, reconstruct the attacks, and take action to stop them.

The table below goes into some depth about IT security activities. You can skip it for now if you like; we will be covering the details later, especially in Chapter 5.

	Tactical Level	Operational Level	Strategic Level
IT Roles	Network Operations Center (NOC) Infrastructure Operations Security Operations Center (SOC)	Incident Response (IR) Team Security Forensics Fraud Detection	Chief Information Security Officer (CISO) IT Management
Tasks	Feed indicators to security products Patch vulnerable systems Monitor, escalate alerts (triage)	Determine details of attacks Remediate Hunt for additional breaches	Allocate resources Communicate with executive management
Problems	Unverified indicators cause false positives Difficult to prioritize patches Too many alerts to investigate	Time-consuming to reconstruct attacks from initial indicators Difficult to identify damage and additional breaches	No clear priorities for investment Executives do not understand technical issues
Value of Cyber Threat Intelligence	Validate and prioritize indicators Prioritize patches Prioritize alerts	Provide "context" to reconstruct attacks quickly Provide data to identify damage & related breaches	Provide priorities based on business risks and likely attacks "Put a face" on adversaries and threats

Table 1-1: Typical problems defending against cyberattacks and how cyber threat intelligence helps address them

Strategic level

At the strategic level, CISOs and IT managers don't have the information needed to set priorities or make budgeting and staffing decisions.

Executives also need information on what not to fear. Today, IT and business managers alike are bombarded with an endless list of potential threats, along with hyperbolic commentary about breaches from vendors and the press. They need information that counters FUD (fear, uncertainty, and doubt) so everyone can focus on the real risks to the enterprise.

Drowning in alerts

A study by the Ponemon Institute quantifies just how hard it is to follow up on security alerts. According to "The Cost of Malware Containment":

- The average organization in the study received 16,937 alerts a week.
- Only 3,218 of the alerts (19 percent) were deemed to be "reliable."
- Only 705 (4 percent) of the alerts were investigated.
- The typical organization spent $1.27 million per year responding to erroneous alerts.

News reports of high-end data breaches point to the same conclusion. At one large retail chain, hacker activity aimed at credit card processing systems generated almost 60,000 alerts over three and a half months. Why didn't the company respond sooner? According to a spokesperson: "These 60,000 entries...would have been on average around 1 percent or less of the daily entries on these endpoint protection logs, which have tens of thousands of entries every day."

Clearly the retailer should have followed up on one of those 60,000 alerts, but we can sympathize with the challenge of deciding which of six million log entries to pursue.

Let's turn our attention now to defining cyber threat intelligence and outlining how it helps organizations defend themselves against targeted attacks.

Cyber Threat Intelligence Defined

The groundswell of interest in cyber threat intelligence derives from the recognition that it is impossible to stop technically advanced adversaries without foreknowledge of their intentions and methods.

Which leads to our definition:

"Cyber threat intelligence is knowledge about adversaries and their motivations, intentions, and methods that is collected, analyzed, and disseminated in ways that help security and business staff at all levels protect the critical assets of the enterprise."

Like all one-sentence definitions of complex ideas, that sounds somewhat abstract. So let's go into a bit more depth and examine some of the key characteristics of cyber threat intelligence as practiced by the leading experts today.

Key Characteristics

Adversary based

The types of intelligence we encounter in books, movies, and news reports focus on specific adversaries. Military and political intelligence activities are directed at enemies of the nation. Law enforcement and anti-terrorism intelligence programs probe criminal gangs and terrorist organizations. Sports teams scout upcoming opponents. Competitive analysts compile information on the products, pricing, and plans of rival businesses.

Cyber threat intelligence activities are also organized around specific adversaries, especially cybercriminals, cyber espionage agents, and *hacktivists*. The enterprise that knows its opponents can optimize its defenses to protect against those adversaries and the attacks they employ.

Risk focused

Cyber threat intelligence programs are based on an assessment of the information assets that the enterprise needs to protect. These assets include data, documents, and intellectual property (such as customer databases and engineering drawings), and computing resources (such as websites, applications, source code, and networks).

Process oriented

From spying, to law enforcement, to competitive analysis, all successful intelligence programs follow the same basic process (Figure 1-1).

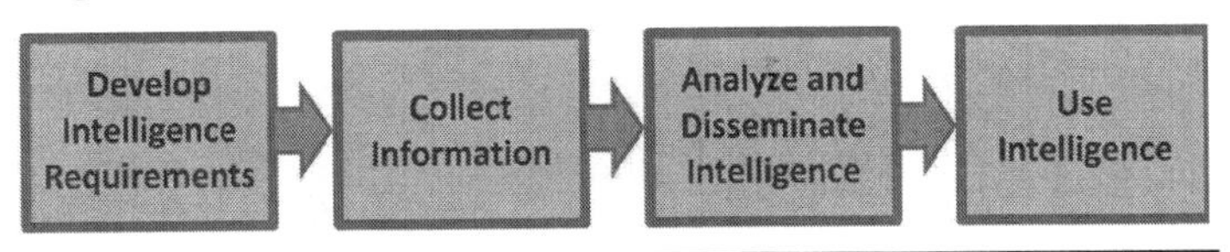

Figure 1-1: The steps in an intelligence process

Don't assume that your organization is following this process informally. Most enterprises never think systematically about what intelligence is required, what sources are available, or how to package information to make it readily usable by different audiences. Make this process explicit and document the tasks at each step. The next four chapters of this guide will help you by outlining many of those tasks.

Tailored for diverse consumers

Another key characteristic of cyber threat intelligence is that it does not stop at distributing raw threat data. Data and analysis must be tailored for each type of intelligence consumer. For example, in respect to the same alert:

- ☑ SOC analysts may want just enough context to know if the alert is worth escalating to the IR team.
- ☑ The IR team may want very detailed context to determine if the alert is related to other events observed on the network.
- ☑ The CISO might want an evaluation of the risk to the organization and a summary connecting the alert to data breaches recently reported in the press.

The Benefits of Cyber Threat Intelligence

Benefits of cyber threat intelligence at the tactical level include:

- ☑ Removing invalid indicators so they don't create false positives
- ☑ Prioritizing patches so the most dangerous vulnerabilities can be fixed first
- ☑ Automating the flow of valid information to SIEMs so they can correlate events with attacks more quickly and accurately
- ☑ Prioritizing indicators so SOC analysts can rapidly identify alerts that need to be escalated

Benefits at the operational level include:

- ☑ Providing situational awareness and context so IR teams can expand their investigations from individual indicators to determine attackers' intentions, methods, and targets
- ☑ Allowing IR and forensics teams to quickly remediate damage done by breaches and prevent additional attacks in the future

Benefits at the strategic level include:

- ☑ Providing managers with an understanding of actual threats to the business (which are different from those hyped by the press) so they can allocate budget and staff to protect the most critical assets and business processes
- ☑ Helping CISOs communicate with top executives and board members about risks to the business, the probable actions of adversaries in the future, and the return on investments in security

Helping management decide how to budget to adequately mitigate risk is one of the most important uses of cyber threat intelligence. Never look at intelligence as a resource for IT security professionals only.

Chapter 2

Developing Cyber Threat Intelligence Requirements

In this chapter

- Review the information assets that must be protected
- Examine the value of identifying adversaries and intelligence consumers

"If you don't know where you are going, you'll end up someplace else."

– Yogi Berra

Cyber threat intelligence requirements guide not only what intelligence is collected, but also how it is analyzed and used. Developing a good set of requirements helps the security organization:

- ☑ Monitor the right threat actors
- ☑ Collect the most useful intelligence
- ☑ Prepare intelligence in the right format and level of detail for each type of user
- ☑ Avoid wasting time and money collecting and disseminating trivial data

Assets That Must Be Prioritized

Fredrick the Great said about war: "He who defends everything, defends nothing." That maxim applies equally to cybersecurity. No security operations center (SOC) can monitor every application, network segment, system, and data store. No incident response (IR) team can follow up on every alert and security event. No manager can budget for every new security technology that comes along.

Let's review the main types of information assets that need to be considered, and the risks to the business if they are lost.

Cast a wide net when you think about high-value assets. Some are not obvious at first. What might happen if bad guys get their hands on your draft press releases? What about project bids and executive emails? Could your CFO's Facebook posts provide ammunition for a *spear phishing* attack?

Besides enumerating information assets that could be attacked, try to quantify the potential costs if attacks on these assets are successful.

Credit card and financial account data

Credit card numbers, bank account numbers, and account access credentials are extremely valuable to cybercriminals because they can be sold in bulk on underground websites.

The costs of losing such data are extremely high. They include data breach notification costs, expenses to provide customers with credit monitoring services, legal fees, regulatory fines (especially for violations of the Payment Card Industry Data Security Standard and banking industry regulations), and revenue declines due to lost customer confidence.

Personal information

Personally identifiable information (PII) includes names, addresses, birthdays, Social Security and national identification numbers, and medical records. PII can be employed in mass phishing attacks, sold on underground websites, and used to create fraudulent accounts that criminals can monetize. Cybercriminals and hackers can also use it as the basis for spear phishing attacks against target enterprises.

The costs of losing PII include breach notification costs, regulatory fines (based on regulations like HIPAA and a host of data privacy laws across the world), legal fees, lost customer confidence, and an increased vulnerability to spear phishing attacks.

Today a great deal of personal information is exposed on social media. Cybercriminals sometimes research executives on Facebook, LinkedIn, and other sites. They use personal details acquired there, such as membership in industry, social, and civic organizations, to create credible phishing email messages. You should educate your employees on the risks of posting excessive information on social media.

Intellectual property

Intellectual property (IP) includes engineering designs, software programs, product manuals, technical documents, and creative works such as videos, music recordings, and books. Theft of an enterprise's IP can mean the loss of competitive advantages. Theft of IP entrusted to you by someone else can result in violations of license agreements and contractual obligations.

Confidential business information

Confidential business information includes business plans, customer lists, competitive bid information, and trade secrets. Their loss can result in a diminished competitive position.

Confidential business data also includes inside information on financial results, mergers, and other news that affects stock prices. The leak of confidential memos, emails, and draft press releases on those topics can prove costly to stockholders, and even trigger criminal investigations.

Credentials and IT systems information

Login credentials and IT systems information can be extremely valuable to adversaries, potentially opening the way for the loss of every type of information asset in the enterprise.

Be sure that suppliers, service providers, and other third parties with access to your systems are diligent in protecting their login credentials for your systems. Some of the most damaging data breaches in recent years started when user IDs and passwords were stolen from third parties.

Operational systems

Operational systems are not assets in the usual sense of the word. However, *distributed denial of service (DDoS)* attacks that bring down corporate websites, and malware that disables business processes, can have a significant impact on revenue, productivity, and public image.

Adversaries

The second part of developing cyber threat intelligence requirements is determining which adversaries might target your enterprise. This analysis can help you decide:

- ☑ Which categories of threat actors to monitor
- ☑ The types of threats that should be given first priority in your monitoring and incident response activities
- ☑ Which adversaries and threat types do not require a significant investment of resources (this is sometimes called "shrinking the problem")

Table 2-1 summarizes the most important adversary types.

	Cybercriminals	Competitors and Cyber Espionage Agents	Hacktivists
Motivation	Obtain financial returns	Obtain commercial, political, or military advantages	Express political beliefs and ideologies Discredit or damage opponents
Assets Targeted	Credit card and financial account data Personal information Credentials	Intellectual property Business information Credentials	Operational systems Credentials
Attack Types and Tools	Malware Phishing Social engineering Botnets Credential escalation Many others	Malware Phishing Social engineering Botnets Credential escalation Many others	Malware Phishing Social engineering DDoS

Table 2-1: Types of adversaries, their targets, and their weapons

Cybercriminals

Cybercriminals are motivated by financial gain, and typically target financial account and personal information that can be converted directly to cash or "monetized" (i.e., sold to criminals on underground websites).

Of course, the term "cybercriminal" covers a multitude of threat actors with very different goals and skills. Some target one industry, such as retail, financial services, local government, healthcare, or media. Others attack specific applications or systems, such as customer databases, human resource applications, or retail point of sale (POS) systems.

You should view the identification of relevant adversaries as an ongoing process. You need to keep up with new cybercriminal types and other threat actors as they emerge, and with new attack types and tools as they evolve.

Competitors and cyber espionage agents

Cyber espionage involves stealing confidential information to obtain commercial, economic, political, or military advantages.

Cyber espionage has long been familiar to military organizations, aerospace and defense companies, and federal government agencies. Now it is being detected by an ever-widening circle of companies that bump up against foreign competitors.

Cyber espionage is carried out by commercial companies, by government-sponsored agents on behalf of commercial companies, and by government and military organizations. They target a wide range of IP and confidential information that can be used to shortcut product development, win competitive bids, and anticipate business strategies, or to gain advantages in military or political struggles.

Be sure to involve line of business managers in your assessment of what IP and business information competitors and cyber espionage agents might target. IT professionals may be unaware of the value (or even the existence) of critical documents, designs, and plans. And remember that software programs are often a key source of competitive advantage!

Hacktivists

Hacktivists attempt to carry out disruptive actions to express their political, social, or ideological beliefs, or to discredit or damage representatives of opposing views. They range from individuals, to loosely connected groups, to well-funded proxies for governments and military forces. In many cases their desire for publicity leads them to be more openly destructive than other types of threat actors.

Unfortunately, few enterprises are immune today. Banks, restaurant chains, retailers, media outlets, social networking companies, and many others are being targeted as agents of capitalism, promoters of disliked cultural values, or symbols of their home government.

Anticipate "trigger actions" that might cause hacktivists to target your enterprise. Are you announcing layoffs or expansion into new geographic markets? Are you involved in environmental or legal controversies? Are you doing business with dissidents or with enemies of repressive governments? Establish a communications process with business managers so they are aware of the risks and you are not blindsided by unexpected business decisions.

Intelligence Consumers

To develop cyber threat intelligence requirements, you must also understand the needs of the people and systems using the intelligence. Those needs include both the information content people require to do their jobs and the formats that make information accessible to people and security systems.

We look at some of these requirements here. In Chapter 5 we explore in more depth exactly how different consumers use cyber threat intelligence.

Tactical users

Network operations center (NOC) staff members need valid malware signatures and URL reputations to allow firewalls, malware gateways, IDS/IPS systems, and other gateway security products to stop attacks without blocking legitimate traffic or generating false positives.

Infrastructure groups that manage servers and endpoint devices want intelligence about which vulnerabilities are most critical for the enterprise so they can decide which security patches to apply first, and which systems should have priority for patches and configuration updates.

SOC analysts monitor alerts and decide which ones should be escalated for further analysis. They want relevant, accurate, and timely data fed to their SIEMs, as well as basic context for alerts so they can quickly decide which ones are isolated events and which might be part of complex attacks.

Operational users

Operational users of intelligence, such as IR teams, forensic analysts, and fraud detection departments, need detailed context around alerts and events. They also need in-depth intelligence on attacks and adversaries so they can:

- ☑ Quickly establish if alerts or events are part of complex attacks
- ☑ Expand their investigations to identify other elements in the attacks
- ☑ Identify the sources of attacks (a process called *attribution*)
- ☑ Determine which systems have been compromised, and which systems need to be remediated

The types of intelligence they need for these activities include analyses of malware, breakdowns of targeted attacks, and reports on the *tactics, techniques, and procedures (TTPs)* of specific adversaries.

Strategic users

Strategic users, including CISOs and IT managers, want threat intelligence reports that enable them to understand trends and make better decisions about security budgets, process improvements, new technologies, and staffing levels. Good intelligence helps them minimize risks and protect new business and technology initiatives.

Chapter 3

Collecting Cyber Threat Information

In this chapter

- Explore three types of threat information
- Find out where each type of information can be collected

"'Data! Data! Data!' he cried impatiently. 'I can't make bricks without clay.'"

– Sherlock Holmes

Information is not intelligence, but it is the raw material out of which intelligence is produced through analysis.

Enterprises today have access to literally terabytes of cyber threat information in the form of huge databases of logs, malware signatures, and other indicators of compromise. Yet most IT groups fail to take advantage of the full range of information sources available to them.

This chapter provides an overview of cyber threat information types grouped into three categories, as shown in Figure 3-1. We also discuss where the information can be obtained.

	Threat Indicators	Threat Data Feeds	Strategic Cyber Threat Intelligence
Content	File hashes and reputation data	Statistics, trends, survey data, and analyses of malware	Information on adversaries and their motivations, intentions, tactics, techniques, and procedures
Key Uses	Increase the effectiveness of blocking technologies and generate alerts	Help SOC and IR teams identify patterns associated with attacks	Help IR and forensics teams analyze attacks, hunt for breaches, and remediate; help managers improve defenses and invest strategically
Primary Sources	Honeypots and scanners on networks	Statistical analyses of indicators, surveys, and sandboxing products	Hacker web forums, underground marketplaces, and personal contacts

Table 3-1: Categories of threat information

Level 1: Threat Indicators

A threat indicator, or indicator of compromise (IOC), is an entity that indicates the possibility of an attack or compromise of some kind. The most common types are file hashes (signatures), and reputation data on domains and IP addresses that have been associated with attacks.

File hashes and reputation data

A malware file hash is a unique identifier of a specific virus, worm, Trojan, rootkit, keylogger, or other type of malicious code. A file known to contain malware is run through an algorithm, most frequently MD5 or SHA-1, which creates a unique "fingerprint" based on the sequence of bytes in the file. The result is a text string that looks like this: 15901ddbccc5e9e0579fc5b42f754fe8.

Domain, IP address, and URL reputations are risk ratings of computers and web pages on the Internet. High risk scores are assigned to websites and systems associated with:

- ☑ Malware and spyware
- ☑ Spam
- ☑ Phishing and other frauds
- ☑ P2P networking and anonymous proxy tools
- ☑ *Command and control (C&C)* servers that manage botnets
- ☑ Data exfiltration servers
- ☑ IP addresses that cannot be traced (the "darknet")

Reputation scores can also be assigned to computers and websites that have been compromised, even if they are not completely under the control of a malicious actor.

Most indicators are merely that – indicators. Some can be regarded as proof of malicious activity, say the hash of the Conficker worm or the domain of a website used exclusively by hackers. Many others, however, indicate only the possibility of an attack. For example, hackers might plant disguised malware on a popular app download site; files coming from this site should be flagged, but not every app from the site is infected.

Technical sources: honeypots and scanners

Finding malware

To create malware signatures, cybersecurity researchers first need to find files infected by malware circulating in the wild. They do this by creating networks of *honeypots*, computers that simulate the activities of web servers, email servers, and other systems, and of computer users surfing the web. These sensors collect files and emails that would be encountered by corporate systems and users during the course of normal operations.

Researchers take files that have not been seen before and subject them to static analysis (examining the code to uncover program instructions and text strings associated with malware) and to dynamic or behavioral analysis (allowing the code to execute and observing malicious actions). If either type of analysis indicates that a file contains malware, the researchers create a signature.

Determining domain and IP address reputations

Researchers extract URLs from web pages and emails collected by honeypots. They investigate to see if the source domains and websites appear to be under the control of threat actors, or have been compromised by malware.

They also analyze emails found by the honeypots to see if they contain indicators of spam, phishing attacks, or fraud. Clues include irregularities in the email header, certain keywords and phrases, and links to known spam and phishing sites.

Researchers can also utilize scanner programs that "surf the web" and test accessible servers for signs of compromise and malicious activities.

The researchers use the results of these analyses to assign reputation or risk scores to domains, IP addresses, and URLs.

Industry sources: malware and reputation feeds

Very few enterprises have the resources to maintain their own threat research groups. Instead, they obtain malware signatures and domain reputation data from a variety of sources, including:

- ☑ Cybersecurity vendors, including antivirus vendors and domain reputation services
- ☑ Cyber threat intelligence firms
- ☑ Independent cybersecurity labs and researchers
- ☑ Open source cybersecurity projects, malware and spam clearinghouses, and real-time blacklist (RBL) providers
- ☑ Government and industry groups that share threat data

Some cybersecurity vendors and cyber threat intelligence firms consolidate contributions from multiple sources. That makes it easier for individual enterprises to obtain a very wide range of data from one signature and reputation feed.

For lists of malware, connect to StopBadware at: https://www.stopbadware.org/clearinghouse or to WildList at: http://www.wildlist.org/CurrentList.txt

For a variety of domain and URL blacklists, connect to the Spamhaus project at: https://www.spamhaus.org/ or to OpenBL.org at: http://www.openbl.org/lists.html

Level 2: Threat Data Feeds

Threat data feeds provide information that correlates and analyzes threat indicators. They help security teams identify patterns associated with attacks. We also include in this category malware analyses that help incident response (IR) teams understand the behavior of malicious software files.

Cyber threat statistics, reports, and surveys

Statistics, reports, and surveys help security teams focus on the most prevalent attacks and alert them to emerging threats.

Statistics

Industry organizations and cybersecurity vendors provide statistics on malware, spam, botnets, and other elements of cyberattacks.

For statistics on malware, try the website of your antivirus vendor, or connect to AV-TEST Institute at: http://www.av-test.org/en/statistics/malware/

For statistics on spam, connect to AV-TEST Institute at: http://www.av-test.org/en/statistics/spam/

For statistics on phishing, connect to the Anti-Phishing Working Group at: http://apwg.org/resources/apwg-reports/

Reports and surveys

A number of vendors and consulting firms publish detailed

reports on various aspects of cyber threats and cybersecurity. These reports typically include:

- ☑ Statistics and trends on different types of attacks
- ☑ Survey results (typically capturing the experience and opinions of that fount of wisdom, the "IT decision maker")
- ☑ Analysis from experts

Survey data can be useful because it gives a picture of the experiences, successes, and failures of IT organizations in responding to threats, and because it can be used to benchmark an individual organization.

The Verizon Data Breach Investigations Report (DBIR) presents analysis of many common threats, and statistics on how well (or poorly) enterprises are responding. To obtain a copy, connect to: http://www.verizonenterprise.com/DBIR/

The Ponemon Institute reports provide invaluable data on the cost of data breaches. To obtain copies, connect to: http://www.ponemon.org/library/

Other useful reports include the Microsoft Security Intelligence Report (http://www.microsoft.com/sir), the Cisco Annual Security Report (http://www.cisco.com/web/offers/lp/2015-annual-security-report/index.html), and the Symantec Internet Security Threat Report (http://www.symantec.com/security_response/publications/threatreport.jsp)

Don't miss our survey of over 800 IT security decision makers and practitioners, which provides a 360-degree view of organizations' security threats, current defenses, and planned investments. For the CyberEdge Group Cyber Threat Defense Report, connect to: http://www.cyber-edge.com/2015-cdr

Surveys are useful, but sometimes they need to be taken with a grain of salt. Results can be affected by sample bias (when the organizations surveyed are different from yours in size, location, or industry) and response bias (when respondents don't want to admit how bad things are).

Malware analysis

Malware analysis provides valuable insights into the behavior of malware samples and the intentions of the attackers behind them.

The most-detailed automated malware analysis is provided by dynamic analysis or *sandboxing* technology. With sandboxing, a suspicious file is allowed to run in a virtual execution environment isolated from the corporate network. The sandboxing product observes and documents all of the actions taken by the file, including malicious activities such as:

- Making unusual entries to the registry
- Disabling antivirus software on the system
- Searching on the network for files whose names include "admin" or "password"
- Making callouts to command and control servers on the web
- Connecting to servers used to stage and exfiltrate stolen data

The observed behaviors not only show whether the sample is malicious, but also provide evidence about the attacker's goals and methods.

You can't rely entirely on sandboxing to identify unknown malware. Many of today's advanced malware files evade sandboxing technologies by executing only if they detect human activities such as mouse clicks, or by verifying that they are on a standalone desktop or server and not in a virtual environment (which is typical of sandboxing products).

Level 3: Strategic Cyber Threat Intelligence

Strategic cyber threat intelligence is information about the specific adversaries targeting your enterprise and the dangers they pose in the immediate future.

Monitoring the underground

The cybercriminals, cyber espionage agents, and hacktivists we have been discussing have developed an entire underground universe where participants:

- ☑ Exchange ideas about targets, tactics, tools, and other facets of cybercrime, cyber espionage, and hacktivism
- ☑ Share expertise on creating and using malware, exploits, spear phishing campaigns, DDoS attacks, and other malicious tools and techniques
- ☑ Plan and coordinate ideologically and politically inspired attacks and campaigns
- ☑ Buy and sell exploit kits, weaponized exploits, obfuscation and evasion tools, and other cyber attack tools
- ☑ Provide services to other threat actors, ranging from specialized tasks (fake website design, password cracking) to outsourcing of infrastructure and complex activities (hackers for hire, rent-a-botnet, DDoS-as-a-service)
- ☑ Buy and sell digital assets, including credit card and Social Security numbers, personal information, and login credentials

The media for these exchanges include online forums, email, instant messaging platforms, social media, and even full-featured online stores.

While most of these venues are open to the public, some of the most important operate on an invitation-only basis and are very hard to crack for outsiders.

If you want to visit these sites, brush up on your language skills. According to the RAND Corporation, the majority of underground forums conduct business in languages other than English, such as Russian, Ukrainian, Mandarin, German, and Vietnamese. Also, the most sophisticated groups practice good operational security, so it may require significant effort to build a cover identity and gain access.

For an eye-opening report on the extent of the cybercrime underground, download a copy of the RAND Corporation study, Markets for Cybercrime Tools and Stolen Data: Hackers' Bazaar, at: http://www.rand.org/pubs/research_reports/RR610.html

Motivation and intentions

Researchers can collect a wide variety of information in this online underground, starting with motivation and intentions. Motivation and intentions provide evidence of which adversaries are likely to attack your industry and your enterprise, and which of your assets they are most likely to target.

The motive of cybercriminals is usually obvious: to make a profit. But their intentions can vary. Some target a certain kind of financial or personal data; others focus on a specific industry.

Competitors and cyber espionage agents exhibit a wider variety of motivations and intentions. These include stealing product designs, intellectual property, and business plans, uncovering the details of bids and proposals, and obtaining political and defense-related intelligence.

Hacktivists display the widest range of motivations, from impressing friends, to advancing a cause such as environmentalism, to discrediting individuals or companies with opposing views, to harassing opponents of a government. They may even aim to shut down part of an economy or national infrastructure in the event of a political conflict. Intentions can be equally varied, including stealing information that can prove embarrassing, defacing or disabling websites, taking over social media accounts, and shutting down crucial services.

Monitoring underground forums can also produce information on threat actors' immediate plans. Some hacktivists announce their upcoming actions online, either to promote their ideology or to coordinate the activities of like-minded individuals and groups.

Although cybercriminals and cyber espionage agents are more secretive than hacktivists, sometimes it is possible to anticipate their actions by looking at information they share. Also, analyzing the malware and services they trade in underground marketplaces can disclose their intentions, targets, and techniques—provided you are able to penetrate their forums.

Tactics, techniques, and procedures

Foreknowledge about adversaries' tactics, techniques, and procedures (TTPs) is extremely valuable. It not only helps enterprises learn what to look for to detect attacks, it guides them on where to strengthen security technology, staffing, and processes.

Researchers can often deduce a great deal about adversaries' TTPs by watching their activities on the web. Valuable evidence includes:

- ☑ Discussions of plans and tactics on forums and social media sites
- ☑ Exchanges of information about new exploits and tools being developed
- ☑ Purchases of tools and services
- ☑ Behaviors of malware and tools offered for sale
- ☑ Sale of credit card numbers, personal information, and other digital assets

Now that we have looked at the kinds of cyber threat information that researchers can collect, we turn our attention to how they convert that information into useful intelligence.

Chapter 4

Analyzing and Disseminating Cyber Threat Intelligence

In this chapter

- Review the difference between information and intelligence
- Examine methods to validate and prioritize threat information
- Explore requirements for analyzing, customizing, and disseminating intelligence

"A person who is gifted sees the essential point and leaves the rest as surplus."

– Thomas Carlyle

Information by itself has limited value. Actionable intelligence must be timely, accurate, and relevant (Figure 4-1). This chapter looks at what is involved in converting information into actionable cyber threat intelligence.

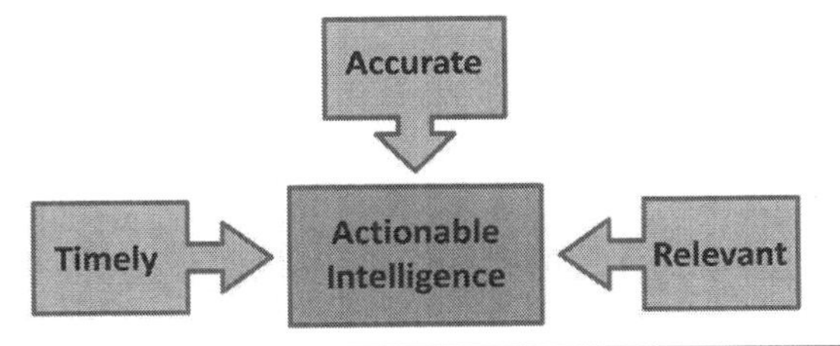

Figure 4-1: Requirements for actionable intelligence

Information versus Intelligence

At the time they are collected, most threat indicators are:

- ☑ Unvalidated and not prioritized
- ☑ Isolated and without context
- ☑ Generic, in the sense of not being associated with any particular type of enterprise

Relying on this kind of information creates serious issues:

- ☑ Security operations center (SOC) analysts can be overwhelmed by alerts and unable to identify the important ones (remember the survey we mention in Chapter 1, where only 19 percent of alerts were reliable and only 4 percent could be investigated).
- ☑ Incident response (IR) teams cannot relate individual alerts to specific threat actors or attack campaigns without laborious, time-consuming research.

Intelligence is information that has been validated and prioritized, connected to specific actors and attacks, customized for specific enterprises, and tailored for specific security consumers within the enterprise.

One of the key goals of cyber threat intelligence is to reduce the amount of time wasted chasing low-level threats, attacks aimed at companies in other industries, and exploits targeting applications and systems not present in your enterprise.

Validation and Prioritization

Validation and prioritization are important because a high percentage of threat indicators are redundant, out of date, or related to threats such as spam and spyware that are typically a very low priority for enterprise security.

In addition, security groups want to give priority to the threats that are most relevant to their own industry, location, applications, regulatory environment, and risk profile.

Risk scores

One method of providing validation and adding context to threat indicators is to attach risk scores. Risk scores can be used by SIEMs and other automated tools to classify and rank alerts, and by human analysts to make faster decisions about which indicators are important.

Risk scores can be created by:

- ☑ Automated technical analysis (Did sandboxing reveal malicious actions?)
- ☑ Statistical analysis (How often has this indicator been associated with attacks on our kind of business?)
- ☑ Human judgement (How vulnerable are we to this threat, and what is the potential impact on our organization?)

Tags for context

Both automated systems and human analysts can use context to determine if an indicator is an isolated event or part of a complex attack. Context can be provided by adding threat and technical tags to indicators. For example:

- ☑ A "Financial" tag could be added to a malware signature associated with an attack on ATM systems.
- ☑ An "Eastern Europe" tag could be added to a URL associated with spear phishing attacks on companies in that region.
- ☑ A "Citadel family" tag could be added to indicators associated with the Citadel credential theft malware family.

With these tags added, a bank's SIEM could be programmed to flag indicators with a "Financial" tag. SOC analysts at a manufacturing company with subsidiaries in Poland and Hungary could give high priority to indicators with the "Eastern Europe" tag. An IR team encountering an indicator with a "Citadel family" tag could immediately retrieve information on the cybercriminal groups and attack campaigns associated with that malware family.

Human assessment

Sometimes there is no substitute for risk assessments by experienced threat analysts. Figure 4-2 shows two examples from a media highlights service. This service comments on threat reports that appear in the press. In the first of these examples the analyst finds the reports accurate but not very relevant. In the second, the threat is important enough to warrant immediate action.

HACKERS SLIP PAST TWO-FACTOR AUTHENTICATION SECURITY AT 30+ BANKS

FROM THE MEDIA

A new campaign that appears to originate in Russia and Romania has compromised customer accounts at 34 banks in Europe and Asia and appears to bypass 2-factor authentication (2FA). The operation discovered jointly by Trend Micro and several affected banks sends victims spear phishing e-mails that reroute users to a malicious page where they download a smartphone app purporting to be a 2FA process.

Read the Story: Seculert

iSIGHT PARTNERS ANALYST COMMENT

While the description of the campaigns appear to be accurate, actors using malware or social engineering to bypass 2FA systems is not a new threat. It has been observed in various forms since at least 2007. Malware capable of stealing 2FA passcodes is widely available in underground markets and is particularly popular in mobile banking malware. The existence of this malware does not negate the security benefit of using 2FA, as each additional security measure increases the barrier of entry for actors to compromise an account.

AGGRESSIVE SELFMITE SMS WORM VARIANT GOES GLOBAL

FROM THE MEDIA

A new version of the Selfmite Android SMS worm is infecting users and allegedly more dangerous than the original. Selfmite.b has been tracked in Canada, China, India and Russia, among other countries. The worm infects users by disguising itself as a link in a text message and then socially engineers victims into clicking on it. Once infected, the worm resends itself to all of the victim's contacts.

Read the Story: **Help Net Security**

iSIGHT PARTNERS ANALYST COMMENT

Selfmite is an "adware" app that earns money for the attackers by encouraging victims to sign up for various subscription services. Although GoDaddy (the owner of the URL-shortening service in its SMS lures) has deactivated the links to the malware, its creators will likely release a new version with different links in the near future. Users concerned about Android malware should avoid downloading apps directly from SMS messages, website links, unofficial appstores or any source other than Google Play. Users should also be wary of Google Play apps posted by unfamiliar developers.

Figure 4-2: Analyst evaluations of media reports about threats

ON THE WEB

You can find selected media highlights reports each week on the iSIGHT Partners blog at: http://www.isightpartners.com/blog/.

Interpretation and Analysis

Let's look at how interpretation and analysis can convert threat information into actionable cyber threat intelligence.

Reports

Threat analyses

What we like to call "anatomy of an attack" reports provide IR teams, forensics analysts, and anti-fraud groups with detailed analyses of specific threats.

The content of threat analysis reports typically includes:

- ☑ Attribution of the attack to specific groups
- ☑ History of the attack and where it has been observed
- ☑ Motives and intentions of the attackers
- ☑ A description of typical victims and targets according to industry, location, vulnerabilities, and other factors
- ☑ An assessment of the impact of the attack and the risks for different types of enterprises
- ☑ A breakdown of the tactics used in the attack, such as reconnaissance steps, phishing and social engineering campaigns, type of malware used, systems compromised, command and control techniques, and data exfiltration methods
- ☑ Similarities to other threats
- ☑ Descriptions of indicators and events that can be used to identify the attack
- ☑ Descriptions of mitigation options and actions that can be taken to protect against the attack
- ☑ The outlook for future appearances of the attack

You can see examples of threat analysis reports at:

http://info.isightpartners.com/the-citadel-banking-malware and http://info.isightpartners.com/newscaster-iranian-threat-within-social-networks

Threat landscapes

A threat landscape report provides a "big picture" view of the threats facing an enterprise. Typical content includes:

- ☑ A review of the business risks facing the enterprise
- ☑ An examination of threat types affecting the enterprise and similar organizations
- ☑ An overview of the adversaries most likely to target the enterprise, including their motivations, intentions, tactics, techniques, and procedures
- ☑ A ranking of security priorities

Analyst skills

Of course, information doesn't magically convert itself into intelligence. The quality of analysis depends on the skills and experience of the people producing it. The individuals who interpret and analyze threat information need:

- ☑ Technical expertise in how malware operates
- ☑ Knowledge of how cybercriminals and hackers construct and execute campaigns
- ☑ Experience with cybersecurity technologies
- ☑ Intelligence *tradecraft* skills on how to uncover and interpret information about threat actors
- ☑ Analytic and critical thinking skills to produce recommendations that are relevant and actionable

People with diverse language skills and cultural backgrounds are also valuable. Many hacker forums use languages other than English and have their own slang, which is difficult for outsiders to interpret.

Communications skills are also important. At least some members of the cyber threat intelligence team need to be able to communicate effectively with a variety of intelligence consumers, including non-technical executives.

Intelligence platform

Threat analysis requires correlating many pieces of information to detect patterns and uncover attacks. Analysts can benefit from an "intelligence platform" to speed up the process. Key elements of such a platform include:

- ☑ A knowledge base to store threat information
- ☑ Automated tools and threat feeds to collect and process human and technical threat data
- ☑ Analytical tools to correlate information
- ☑ Publishing tools to automate the creation and distribution of intelligence "products"

Customization

Customization should be built into the cyber threat intelligence process at two levels.

Customization for the enterprise

Out of the mass of threat information available, an intelligence analyst needs to select the facts most relevant to the specific circumstances of the enterprise. These circumstances include its industry, location, size, regulatory and political environments, business risks, methods of interacting with customers, software applications, and use of mobile technologies and cloud resources.

Customization for the consumer

The intelligence analyst also needs to tailor information and analysis to each type of consumer, including tactical groups who want basic information to make decisions quickly, IR and forensics teams who want as much information as possible to help them identify and assess attacks, and IT managers who prefer summaries.

Often the same information should be reformatted and summarized for different audiences. This may seem like unnecessary effort, but it is a fact of business life that busy people often lack the time to sift through detailed data to find the key points they need.

Dissemination

Different audiences also have different preferences for how often and in what form they receive cyber threat intelligence. Honoring those preferences can be critical for ensuring that intelligence is used effectively by each group or individual.

Automated feeds and APIs

We have mentioned situations where threat intelligence can be acquired from or shared with SIEMs, antimalware products, firewalls, IPSs, and other security technologies. It is faster and more efficient to share data between those systems and the threat intelligence platform or database with automated feeds and *application programming interfaces (APIs).*

Searchable knowledge base

Cyber threat intelligence is cumulative. An indicator or clue received today often needs to be checked against information and events going back months or years. Enterprises need a searchable knowledge base to store historical threat data.

Tailored reports

Most human consumers prefer to receive threat intelligence in structured reports. The length, level of detail, and focus of the reports will vary depending on the responsibilities of the readers. Frequency is also important: some users will want a constant, up-to-the-minute stream of information, while others might prefer a daily summary. Management types might be most receptive to highly summarized information or reports at monthly or quarterly intervals.

It is important to determine these preferences and needs as part of gathering cyber threat intelligence requirements (see Chapter 2).

Consider how intelligence consumers in your organization might want to receive data. Would they read email summaries or a newsletter? Would it help to push out urgent facts as text messages, or use a Twitter account?

Chapter 5

Using Cyber Threat Intelligence

In this chapter

- Explore how cyber threat intelligence is used by different IT security groups
- Understand how to use intelligence to "pivot" from an initial alert to detect and analyze a complex targeted attack

"What enables the wise sovereign and the good general to strike and conquer, and achieve things beyond the reach of ordinary men, is foreknowledge."

– Sun Tzu

So far, we have described how to develop cyber threat intelligence requirements and how to collect, analyze, and disseminate intelligence.

Now we are ready to discuss the payoff for all of that activity: how intelligence helps IT professionals do their jobs better.

To do that we return to the framework introduced in Chapter 1, which looks at the uses of cyber threat intelligence at the tactical, operational, and strategic levels.

IT Operations: Blocking, Patching, and Triage

At the tactical level, cyber threat intelligence improves the effectiveness of blocking technologies, helps infrastructure groups prioritize their patching activities, and allows security operations center (SOC) analysts to quickly and accurately decide which alerts require action.

Network operations: improve blocking

In most enterprises, the network operations center (NOC) staff manages network security technologies, including:

- ☑ Gateway antimalware products
- ☑ Firewalls, next-generation firewalls (NGFW), and application firewalls
- ☑ Intrusion detection and intrusion prevention systems (IDS/IPS)
- ☑ Secure web gateways (SWG)

These products apply rules to thwart malicious activities (e.g., close unneeded ports so they can't be used for reconnaissance or exfiltrating stolen data) and use threat indicators to block malware and network traffic to and from computers known to be controlled by threat actors.

However, when the quality of threat indicators is poor, the NOC staff usually turns off blocking to avoid cutting off legitimate files and traffic.

Cyber threat intelligence, by validating threat indicators such as malware signatures and domain reputations, can reduce false positives and allow NOC staff to use blocking technologies with confidence. Verified indicators also help SOC analysts escalate fewer non-verified alerts.

Provide NOC staff with access to analysis of threat actors and complex attacks. Details about malicious tools and tactics can help them fine-tune the rules used by firewalls, IPSs, and similar systems.

IT infrastructure groups: prioritize patching

Patch management is a major task for groups that manage servers, endpoints, and network and security devices.

Patching is a very time-consuming process. Infrastructure groups often find themselves with a backlog of patches and difficult choices about which to apply first, particularly on notorious Microsoft "patch Tuesdays." Common Vulnerability Scoring System (CVSS) ratings include an unrealistic number of "high" ratings. Also, the generic "critical/important/moderate/low" ratings issued by antivirus vendors are not reliable guides to the importance of patches for specific enterprises.

You can find CVSS ratings in the National Vulnerability Database at: https://nvd.nist.gov/home.cfm

Cyber threat intelligence helps infrastructure groups prioritize patches based on rich information about vulnerabilities. That information can include technical descriptions of vulnerabilities and their effects, how hard they are to exploit, and whether exploit tools are currently available in the wild.

In fact, researchers with strong cyber threat intelligence tradecraft can find conversations on underground websites where threat actors reveal the vulnerabilities they intend to exploit, the techniques they are using to create exploits, and the specific organizations they plan to target.

By using this intelligence to improve patching priorities, infrastructure groups can close the window on immediate threats faster and spend less time on vulnerabilities that are low priority or irrelevant to their organization.

Pick the patching cycle

One U.S. government agency uses cyber threat intelligence to assign patches to each of its three patching cycles. The high-priority cycle patches all affected systems within 24 hours, the medium-priority cycle within one week, and the low-priority cycle within a month.

SOC: triage for alerts

In many enterprises the SOC analysts review SIEM alerts and divide them into categories such as:

- ☑ Escalate immediately to the incident response (IR) team
- ☑ Investigate when time permits
- ☑ Ignore

Unfortunately, making this determination is difficult because most enterprises generate far more alerts than the SOC and IR teams can investigate.

Cyber threat intelligence can enhance event prioritization and situational awareness in two ways:

- ☑ By attaching risk scores or tags to threat indicators so the SIEM can flag appropriate alerts as high priority
- ☑ By allowing the SIEM or the analyst to query the threat intelligence knowledge base and correlate alerts with additional context about attacks

The staff at a manufacturer, for example, could instruct a SIEM to flag malware associated with attacks on supervisory control and data acquisition (SCADA) systems and process control systems (PCS). The SIEM could also be programmed to send a query to the intelligence knowledge base to automatically return contextual information about the malware. This information might include the adversary behind the attack and whether the malware has been used to target other manufacturers.

If you can program a SIEM to flag high-priority alerts based on tags and enterprise-specific rules, and to quickly assemble related threat information, you will automate some of the most time-consuming tasks required of SOC analysts. These capabilities enable analysts to quickly and accurately determine which alerts to escalate, and to be more productive in managing threat indicators.

Incident Response: Fast Reaction and Remediation

At the operational level, cyber threat intelligence helps IR teams, as well as forensics, security analysis, and fraud detection groups, analyze complex attacks more quickly and more thoroughly.

Accelerating attack analysis

When an attack is detected, the IR team needs to answer questions such as:

- ☑ Who is behind the attack?
- ☑ What tactics are they using in their campaign?
- ☑ What information assets are they targeting?
- ☑ How far has the attack progressed, what systems have been compromised, and what data has been accessed?
- ☑ What steps can halt the attack, and then remediate the effects?

Yet at first, the IR team has nothing to go on but the indicator that triggered the initial alarm. Often this is no more than a single malware sample, or a link to a known command and control (C&C) server. It may take an analyst days to put the pieces together by searching through emails, application logs, network traffic, system configurations, threat analyses, and other disparate data sources.

Not only do long research efforts waste the time of expert staff members, they also give attackers more time to find and exfiltrate data and perform other hostile acts.

Cyber threat intelligence can accelerate incident response by providing rich context around an initial indicator. The intelligence knowledge base can quickly answer questions about the indicator, such as its technical characteristics, its effects, and where it been observed in the past.

The IR team can also query the intelligence knowledge base for more information, such as which adversaries have used this technique, what those adversaries target, and which infrastructure and tools they use. This information allows the IR team to *pivot* from the initial indicator to quickly find the related detail they need.

Rich context not only accelerates incident response, it also enables IR teams to detect attacks they might otherwise miss completely. That's because context helps analysts recognize that seemingly isolated events are actually part of a multi-part campaign.

How to pivot and nail an attack

Cyber threat intelligence allows incident response and "hunt" teams to progress from an initial clue to contextual information that helps validate the threat and identify its effects.

Let's say a malware sample is discovered on the laptop of an employee in finance. Is that infection an isolated incident, or part of a broader threat? What should be done to remediate possible damage from the attack?

You can pivot from the initial clue (the malware signature) to investigate related indicators. For example, if the intelligence knowledge base shows that the malware is associated with a specific adversary whose method is to:

- Launch a phishing campaign that targets members of the finance department
- Infect systems with malware that seeks higher-level credentials
- Target a specific financial application
- Use a C&C server at a specific IP address

Then you can check to see if:

- Others in the finance department have received suspicious emails.
- There has been an increase in the number of accounts with privileged entitlements.
- The application has experienced unusual exports of data.
- Other systems on the network have had traffic to or from suspect IP addresses associated with that adversary.

Assisting investigation and remediation

Cyber threat intelligence can also help security teams uncover the effects of related attacks and determine how to clean up the damage.

To increase their odds of success, adversaries often conduct campaigns that use multiple tools and techniques. The IR team can use knowledge of these tactics to hunt for additional breaches these adversaries might have engineered.

Intelligence helps with remediation, too. Knowledge of the tools and tactics used in attacks can help the IR and forensics teams determine which systems on the network have been compromised, and how. That insight makes it easier to locate and remove the attacker's footprints and to set up defenses to protect against the same and similar tactics in the future.

Alert, pivot, respond: A real example

The following is an excerpt from a blog post published by Tim Armstrong of Resilient Systems. HP ArcSight is a SIEM system from HP. The Resilient Incident Response Platform, or IRP (formerly called Co3), is a collaborative tool for managing the work of IR teams. iSIGHT Partners is a specialist in collecting and delivering cyber threat intelligence.

Incident Response Management with the Resilient Incident Response Platform, HP ArcSight and iSIGHT Partners

We all know that Target-like breaches aren't completely preventable. But does that mean we're doomed and powerless? Not even close. A decisive response effort can dramatically reduce the impact of a breach, potentially stopping attacks in their tracks before sensitive data is lost.

Workflow

A security analyst working in ArcSight identifies an event that needs the attention of the Incident Response team. A Point-of-Sale system has a strange file on it. The MD5 hash appears in the ArcSight console. The analyst right-clicks this event in ArcSight, chooses the "Escalate to Co3/IRP" option and the file's MD5 and other pertinent event information is automatically collected from ArcSight and sent to IRP. IRP then automatically builds a detailed incident response plan based on the specific parameters of this incident and notifies the incident response team.

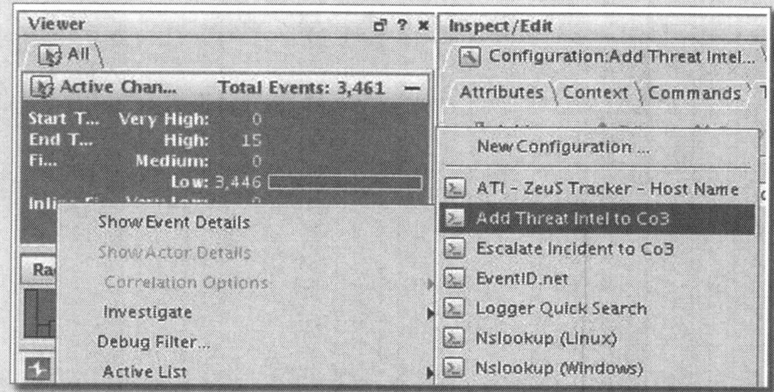

IRP then automatically compares all artifacts against our threat feeds. In this case there is a hit from iSIGHT – it flags the artifact as a sample of the infamous POSRAM malware.

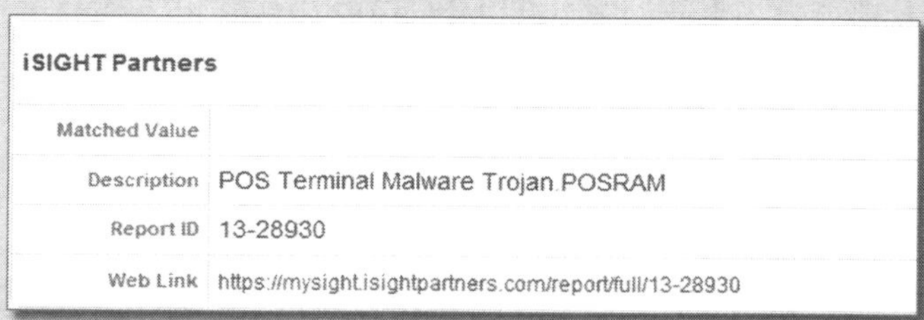

iSIGHT Partners

Matched Value	
Description	POS Terminal Malware Trojan.POSRAM
Report ID	13-28930
Web Link	https://mysight.isightpartners.com/report/full/13-28930

Once escalated, this process is completely automated. The target has gone from an initial indication to an incident determination with deployed IR plan and a team already in motion. Responders in IRP can now assign tasks to the Incident Response group to isolate and contain the threat and drive the incident to closure. The entire incident response process can easily be managed from IRP, following both company and industry best practices.

The average time to respond to serious incidents is currently counted in months. Using Resilient Incident Response Platform, HP ArcSight, and iSIGHT Partners, we've just alerted and responded in minutes, before a security event grew into a company disaster.

Reprinted with permission of Resilient Systems http://www.resilientsystems.com

Management: Strategic Investment and Communications

CISOs and IT managers face difficult decisions about where to invest in new technologies and staff. Of the many threats in the headlines and the many security products touted by vendors, threat intelligence can help identify which ones should be prioritized. Intelligence can also help the CISO and IT managers explain threats in business terms so they can have productive discussions with senior executives and board members.

Investing effectively

Cyber threat intelligence helps IT managers understand challenges such as:

- ☑ New adversaries emerging to target enterprises in their industry
- ☑ New tactics and techniques exploiting weaknesses in current security defenses
- ☑ New "attack surfaces" such as mobile devices, data hosted in the cloud, and employee information posted on social networks

This information allows CISOs and IT managers to invest funds and staff to protect against the most likely attackers and threats, rather than being forced to react to every headline describing a new data breach.

Deprioritization is useful too. Determining which threats are *not* likely to affect a given enterprise leaves more time to pursue the threats that are important.

Improving management communication

IT groups find it difficult to hold meaningful discussions about cybersecurity with executives when they frame issues in terms of technical attacks that should be addressed (APTs, DDoS attacks, social engineering) and technologies that need to be acquired (NGFWs, security analytics platforms, endpoint threat detection).

Cyber threat intelligence helps IT managers put a face on adversaries and explain their motives in human terms: the political activists who want to embarrass the company, the foreign competitor trying to unearth business plans, or the cybercriminal group trying to make money on stolen Social Security numbers.

The same approach also helps IT managers describe security issues in terms of risks to the business, such as the potential loss of revenue from online sales, the impact on regulatory compliance, or the inability to deploy a new mobile application for the sales force this year.

Which report might earn you a bonus?

Top management wants to hear about impact on the business, not technical metrics. Which of these two reports will make you a hero?

Report to the CEO, version 1

Last month we:

- Reviewed 1,452,134 log entries
- Detected 423,132 viruses
- Blocked 2,028,438 connections
- Closed 3,095 incident tickets

Report to the CEO, version 2

Last month we detected and blocked two cybercrime attacks linked to a criminal organization in Eastern Europe that has been targeting POS systems at mid-sized retailers. Our actions:

- Prevented the theft of 10 million customer credit card numbers
- Avoided $78 million in lost revenue and the costs that would have been incurred for notifying customers of the data breach, cleaning up infected systems, and paying regulatory fines and legal fees.

Chapter 6

Implementing an Intelligence Program

In this chapter

- Explore the process for creating a strategic roadmap
- Review best practices for implementing a cyber threat intelligence program

"Victory awaits him who has everything in order."

– Roald Amundsen

This chapter explores eight best practices for implementing a world-class cyber threat intelligence program. Some of the tasks described here are management oriented (develop a strategic roadmap), while others are technical (automate workflows, expand monitoring). The eight tasks and their relationships are illustrated in Figure 6-1.

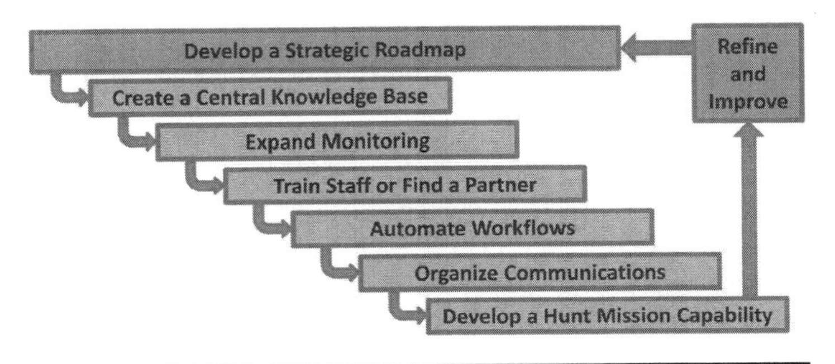

Figure 6-1: Tasks for implementing cyber threat intelligence

Not all of the best practices described in this section need to be perfected right off the bat. But don't skimp on the analysis and planning up front. They will help you fill gaps fast and ensure that your early efforts are directed where they will produce the most benefits.

Develop a Strategic Roadmap

A cybersecurity strategic roadmap is a useful tool for the entire IT organization, and an essential one for the cyber threat intelligence program. It aligns threat intelligence requirements and activities with business risks. Let's take a brief look at some of the key tasks involved in creating a strategic roadmap.

Evaluate assets, adversaries, and defenses

The first step in developing a strategic roadmap is to assess key assets, adversaries, and defenses. Tasks include:

- ☑ Enumerating information assets that need to be protected and assessing the impact of losing them.
- ☑ Identifying IT systems that are critical to business operations, including business applications, public-facing servers, and infrastructure and operational control systems.
- ☑ Identifying likely adversaries and their targets, techniques, tactics, and procedures.
- ☑ Evaluating the effectiveness of current security systems, staffs, and processes, in particular their ability to monitor, detect, mitigate, prevent, and remediate targeted attacks from likely adversaries.

Perform a gap analysis

The next step is to perform a gap analysis that highlights which security systems and processes need improvement. It should focus on gaps that might expose high value information assets. This process requires effort, but it provides a critical advantage by "narrowing the problem." Figure 6-2 illustrates this concept.

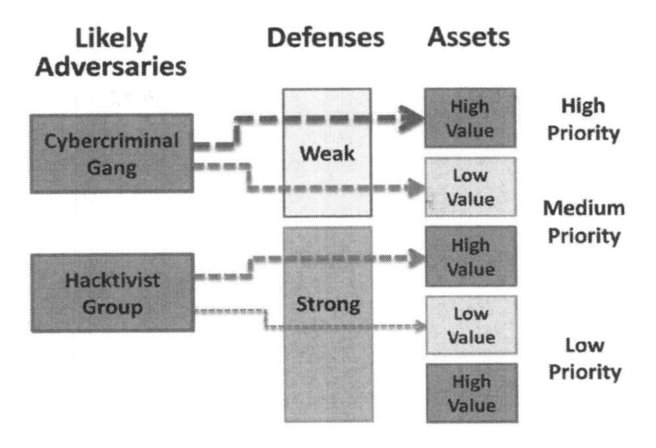

Figure 6-2: A gap analysis helps set priorities

The most important gaps are weak defenses protecting high value assets. Those defenses must be strengthened as soon as possible.

The second priority for improvement is weak defenses protecting low value assets, as well as strong defenses protecting high value assets. These also must be addressed, but not as urgently.

Last on the list are low value assets protected by strong defenses, and assets not targeted by any likely adversary.

Performing a gap analysis is obviously much more complicated than this simple overview suggests, but it is a critical investment for the cyber threat intelligence program. The assessment process gives the organization a sound basis for determining which improvements are most urgent, and also what cyber threat intelligence needs to be collected and analyzed.

The gap analysis may suggest simple but pervasive controls that can mitigate large classes of threats. A good example is removing administrative privileges from users who are likely attack targets. This action can minimize damage from many targeted phishing and malware attacks, because even if the users' credentials are compromised, they can't be used to access as many critical systems or databases.

Outline investment priorities

The final step in developing a security operations roadmap is outlining investment priorities and options. The final report should include a series of recommended improvements with their costs.

An accurate roadmap enables IT managers to conduct productive discussions with executive management and frame recommendations in business terms (see the investment recommendation generator, below). It also provides guidance on priorities for the cyber threat intelligence program.

Handy investment recommendation generator

When you finish your strategic roadmap, create an investment recommendation for each major item by plugging appropriate terms into the spaces indicated.

We recommend **[proposed improvement]** in order to protect **[asset]**. The effects of losing or impairing this asset could result in **[costs, revenue losses, fines, loss of competitive advantage]**. This asset is frequently targeted by **[likely adversaries]** who have previously attacked enterprises similar to ours. Our main defense, the **[security technology or process]**, is inadequate because it lacks **[features, performance, staffing]**. An investment of **[$$,$$$]** will solve **[defects]**, strengthen **[improvements]**, and allow us to mitigate foreseeable risks associated with the **[technical or business initiative]**.

Create a Central Knowledge Base

We now turn to some of the operational steps required to implement an effective cyber threat intelligence program.

A repository or knowledge base is a critical tool. A lot of the work of security operations center (SOC) analysts, incident response (IR) and forensics teams, and others involves correlating data from different sources and time periods.

The information stored should include threat indicators, malware and attack analysis reports, and investigative reports about attacks observed in the enterprise. The information needs to be easily searchable.

Technologies used for the knowledge base could include document management and collaboration systems, SharePoint, databases, and data management tools like Splunk. To get the most out of threat data, it should be possible to integrate the knowledge base with SIEMs and other security systems via an application programming interface (API) or standard connectors (we discuss this more fully in the "Automate Workflows" section, below).

If possible, the knowledge base should include several years of history. Analysts need to reconstruct attacks spanning months or even years. In addition, they should be able to check whether indicators observed today appeared previously.

It is usually a good idea to assign at least a part-time administrator or librarian to the knowledge base. The administrator ensures that the knowledge base is updated properly, that data is organized and normalized so it can be found easily when needed, and that permissions are set so the right data is accessible to the right people.

Expand Monitoring

The more security information that is available to security professionals, the easier it is for them to detect and analyze attacks.

Monitor internal activities and traffic

Organizations must be able to aggregate and correlate log and security event data from servers, security products, and network devices. Typically SIEMs and security analytics tools are deployed for this purpose.

The security organization should also consider placing security monitoring devices at strategic choke points in the network to monitor lateral traffic within the enterprise. Analysts can then use tools such as full-packet capture systems to monitor traffic to and from critical systems. This traffic can reveal attackers attempting to escalate credentials, to communicate with command and control servers, to find confidential data and stage it for exfiltration, or to exfiltrate data to outside servers.

Monitor external threats

The security organization should also be systematic in monitoring external threats through:

- ☑ Threat indicator data feeds
- ☑ News reports and flash analyses of current security events
- ☑ Detailed reports and contextual threat data about malware and attacks
- ☑ Monitoring of adversaries on web forums, underground websites, and hacker marketplaces

Enterprises have the option of outsourcing threat information collection and external threat monitoring to cyber threat intelligence firms, a topic covered in Chapter 7.

Train Staff or Find a Partner

In Chapter 4 we discussed some of the skills and experience required of researchers who collect and analyze cyber threat information. These attributes include extensive technical expertise, knowledge about how attacks are constructed, threat research skills, critical thinking, and a mastery of languages. Not surprisingly, expert cyber threat researchers and analysts are difficult to find in the job market.

It is sometimes possible to train existing security staff for these roles. This process involves finding people with technical expertise and educating them about how to:

- ☑ Explore underground and black market websites
- ☑ Sift clues and correlate data to create coherent pictures of adversaries and their methods
- ☑ Customize and package findings for different intelligence consumers

Another alternative is to find a partner that provides cyber threat intelligence as a service. Chapter 7 outlines criteria for selecting a partner of this kind.

Automate Workflows

Automated workflows ensure that cyber threat intelligence is available to analysts immediately as context around a threat indicator, without significant manual effort.

One option we discussed in Chapter 5 is integrating SIEMs with the threat intelligence knowledge base to support the workflows of the SOC team.

For example, when a SIEM receives an alert, it might automatically query the knowledge base, which would return basic information about the indicator that triggered the alert, together with tags assigned to that indicator. If one of the tags is significant for that enterprise, then the SIEM can flag the alert as high priority.

When a member of the SOC team decides to investigate the flagged alert, the system could also display context about the alert that would help the analyst decide if it should be investigated further, remediated in some way, escalated to the IR team, or ignored. If the alert is escalated, then the context delivered from the intelligence knowledge base could also be forwarded to the IR team.

This process is far more efficient and reliable than forcing SOC analysts to query the knowledge base manually each time they investigate an alert. Integration can be greatly simplified if the knowledge base has an API or a *software development kit (SDK)*.

Organize Communications

Some threat intelligence should be delivered through automated workflows at the moment of need. In other situations, however, the format and timing of communications can be tailored to the preferences of users. This is particularly true for written analyses and reports. For example:

- ☑ SOC and IR teams want intelligence "flashes" on newly emerging threats and adversaries, delivered as soon as information is available, so they can react immediately to zero-day attacks.
- ☑ IR and forensics teams need comprehensive analyses of malware and attacks, provided as soon as all the details are available (these details should also be stored in the knowledge base for retrieval on demand).
- ☑ CISOs and IT managers may prefer summary information about malware and attacks, along with statistics and trend data, as well as reports delivered on a weekly or monthly basis.
- ☑ Executive managers might request quarterly high-level summaries tied to business issues. They might also demand immediate assessments of breaches and security issues when reports appear in the business press. Those help them answer questions from the CEO and members of the board of directors (see Figure 4-2 for examples).

Someone involved in the cyber threat intelligence program should assess communications preferences and set up appropriate processes to reach all members of the IT organization. Tasks might include maintaining distribution lists for emailing or texting time-sensitive information, sending periodic security updates and newsletters, and ensuring that appropriate search tools are available to find information stored in the knowledge base.

Remember the employees. They are often the weakest link in the security chain. Figure out how security information can be packaged and delivered to them in ways that encourage good behavior. How about an internal security newsletter, a Twitter account, or a Facebook page?

Develop a Hunt Mission Capability

Most enterprises use cyber threat intelligence only in a reactive fashion to help respond to alerts and analyze attacks after they have been identified. However, a number of leading-edge security teams have taken a more proactive approach.

The basic idea of a *hunt mission* is to anticipate the most likely threats and aggressively search for indicators that might reveal campaigns and attacks in their earliest stages. This process might involve:

- ☑ Designating specific indicators for special monitoring
- ☑ Tagging these indicators so that when they are encountered by the SIEMs or security sensors on the network, the hunt team is notified immediately
- ☑ Following up immediately by pivoting on the initial indicator and looking for related indicators that confirm an attack is occurring.

When a threat is detected, the hunt team can:

- ☑ See if indicators of the same threat can be found in other parts of the enterprise, perhaps where detection may not have been as thorough

- ☑ Determine whether the same or similar campaigns have been logged in the past
- ☑ Investigate whether the same adversary who launched the attack has used other tactics and techniques, and search for those.

If any of these conditions are true, the hunt team investigates whether a full attack took place, and if so, determines its effects.

Developing a hunt mission capability requires resources, but it has the potential to head off attacks before they cause damage and to identify successful breaches that were not detected by existing security measures.

Refine and Improve

An effective cyber threat intelligence program can never be static. New adversaries are always emerging and coming up with new tactics and techniques. Technologies and business initiatives are always changing. And as your organization gains experience using cyber threat intelligence, improved techniques and processes will suggest themselves.

To keep up with these changes, repeat on a regular basis the analysis performed when the original strategic roadmap was created. This includes assessing new adversaries and attacks, updating the gap analysis, and revising your cyber threat intelligence requirements.

Develop and track metrics whenever possible. You may be able to quantify the quality of alerts, the number of alerts investigated, and the mean time to complete attack investigations. You can also use interviews, structured surveys, and online forums to collect feedback from intelligence consumers. Have them rate on a scale how easy it is to use the intelligece they are getting, and how much it is helping them do their jobs. Ask them how the information and its delivery could it be improved.

Chapter 7

Selecting the Right Cyber Threat Intelligence Partner

In this chapter

- Review different types of cyber threat intelligence partners
- Learn which types to avoid
- Examine criteria for selecting the right partner

"We should not only use the brains we have but all that we can borrow."

– Woodrow Wilson

Only a handful of large commercial firms, government agencies, and military organizations have the resources to handle all aspects of cyber threat intelligence internally.

By now you probably understand why: first-class cyber threat intelligence requires a worldwide network of sensors, a cyber threat research lab, expert researchers and analysts with strong language and communications skills, and a platform to collect and disseminate threat data.

That's why the vast majority of enterprises engage one or more partners to help with information collection and analysis tasks. But what kinds of partners are available? What is the best approach to selecting the right ones for your enterprise?

Types of Partners

Quite a variety of product and service vendors offer some elements of cyber threat intelligence. Roughly speaking, they fall into three categories: companies that focus on threat indicators, companies that combine threat indicators with threat data feeds, and companies that provide comprehensive cyber threat intelligence services.

Providers of threat indicators

A wide variety of security technology vendors and open source projects supply indicators, signatures, and screening rules to power firewalls, antimalware software, IDS/IPS, unified threat management (UTM) systems, and other products. In some cases, the indicators are delivered as raw data; in others, they are accompanied by risk or reputation scores.

Threat indicator feeds are critical at the tactical level for maximizing the effectiveness of blocking technologies. However, they don't provide context for incident response. Unless they are validated, they can waste time by creating false positives and meaningless alerts.

Providers of threat data feeds

A number of technology vendors and security service companies offer threat data feeds. These include collections of indicators that have been validated and prioritized, plus detailed technical analyses of malware samples, botnets, DDoS attack methods, and other malicious tools. They sometimes add statistics and trend information, for example "Top 10" lists, percentage breakdowns of malware types, and locations of spam attacks and botnets.

Threat data feeds help at the tactical and operational levels, giving security operations center (SOC) analysts and incident response (IR) teams diagnostic data about attack tools and basic context about attackers. However, they rarely provide information about the intentions or tactics of adversaries, or help "narrow the problem" to the specific adversaries and attacks most likely to target a specific industry or enterprise.

Providers of comprehensive cyber threat intelligence

A small number of firms offer all three types of threat intelligence: validated threat indicators, threat data feeds, and strategic threat intelligence. The leading companies integrate the three types, for example providing IOCs that have been validated, tagged, and connected to rich context about adversaries. Typical deliverables include:

- ☑ Validated threat indicators with tags
- ☑ Detailed technical analyses of attack tools
- ☑ In-depth research on adversaries, with data collected from underground websites and private sources
- ☑ Detailed studies of existing and emerging threat actors
- ☑ Assessments of threat landscapes facing industries and individual enterprises
- ☑ Assistance developing cyber threat intelligence requirements
- ☑ Threat information customized for different audiences at tactical, operational, and strategic levels.

The rest of this chapter discusses what to avoid and what to seek in selecting a partner to provide comprehensive cyber-threat intelligence services.

What to Avoid

If you are looking for a partner to provide comprehensive cyber threat intelligence services, you should avoid:

- ☑ Security product companies, because their services are almost always designed to support the use of their product, not to optimize overall security
- ☑ Security services companies with a regional focus, because cyber threat information needs to be collected and assessed on a global basis

Some firms provide generic or "one size fits all" threat analyses. These are useful, but they are not a substitute for intelligence that is collected, analyzed, and disseminated based on the specific circumstances of your enterprise.

Important Selection Criteria

Every enterprise needs to develop its own list of criteria for evaluating potential cyber threat intelligence partners, but the following seven factors should be included.

IT analyst firm Gartner has published a research note called *Market Guide for Security Threat Intelligence Services* that discusses criteria for selecting partners, and includes short overviews of many of the firms in that field. If you are not a Gartner client, you can obtain a copy by contacting iSIGHT Partners at info@isightpartners.com.

Global and cultural reach

Find out how many languages the service firm covers, and how many of its researchers and analysts are located on each continent.

Historical data and knowledge

Very few cyberattacks are truly original. Most reuse existing malware, infrastructure, and methods in new combinations, or evolve from older techniques. The same adversaries often attack companies in the same industry repeatedly. For these reasons, several years of historical data and expert experience are invaluable for identifying and analyzing the latest attacks. Ask prospective partners when they created their threat knowledge base, and the average tenure of their threat researchers and analysts.

You can also ask a prospective partner how they maintain the knowledge base and weed out obsolete items.

Range of intelligence deliverables

Make sure your prospective partner offers all three types of threat intelligence: validated threat indicators, threat data feeds, and strategic threat intelligence. They should deliver indicators with tags that both automated systems and people can use to connect IOCs to rich context about adversaries and attacks. A range of intelligence deliverables should be available with the format, level of detail, and delivery frequency appropriate for your key users (e.g. intelligence "flashes" for SOC and IR teams, detailed adversary and threat analyses for IR and forensics teams, and higher-level briefings and trend reports for managers).

APIs and integrations

We have seen how integrating a cyber threat knowledge base with SIEM and other security products can help you prioritize alerts and automate the addition of context to threat indicators. Find out from prospective partners if they support out-of-the-box integration with your SIEM and security products, and if their information delivery systems have an API for creating customized connectors.

Intelligence platform, knowledge base, and portal

Infrastructure is a crucial enabler for a cyber threat intelligence operation. Ask the service firm to:

- ☑ Describe its platform for collecting, analyzing, and disseminating intelligence (or better yet, demonstrate it to you)
- ☑ Explain (or demonstrate) the search and retrieval capabilities of its knowledge base
- ☑ Let you experiment with the customer portal to see how easy it is to use

Client services

We emphasize throughout this guide that cyber threat intelligence should be customized to provide information tailored to the industry, geography, applications, and regulatory environment of each enterprise. Find out how the service firm works with its customers to develop intelligence requirements, to conduct research and perform analysis focused on each organization's adversaries and risks, and to disseminate information tailored to each type of intelligence consumer.

Access to experts

Find out whether your potential partner gives customers direct access to its experts to answer questions, clarify analyses, and exchange ideas. Also, do these experts "wing it" with ad-hoc analyses, or are there formal processes for collecting, analyzing and disseminating intelligence?

On another level, can the analysts and managers at the firm establish credibility with your managers and executives and help justify necessary investments and security initiatives?

Always talk to reference customers. There is no better way to verify the service firm's claims. Besides checking on specific capabilities, remember to ask references about the firm's customer service, responsiveness, and willingness to customize deliverables. Find out if your prospective partner has helped clients make maximum use of cyber threat intelligence across tactical, operational, and strategic levels.

Intelligence-driven Security

We leave you with a final consideration for selecting the right cyber threat intelligence partner.

Many enterprises today are stuck in what can be called "technology-driven security," meaning they are laser-focused on acquiring and implementing the latest security products. A good cyber threat intelligence firm can help those enterprises transition to "intelligence-driven security," guiding them to think strategically about how best to manage risks and invest resources to defeat the most dangerous adversaries.

We contend that only enterprises that have made this transition will be prepared to address the increasingly sophisticated targeted attacks that are destined to emerge over the next few years.

Glossary

advanced persistent threat (APT): A targeted cyberattack that leverages multiple tactics to gain network access and remain undetected for extended periods.

application programming interface (API): A set of documented commands, functions, and protocols that allow software programs to communicate and share data.

attribution: Linking an attack to a specific threat actor.

command and control (C&C) server: A server operated by a threat actor to provide instructions to bots or to communicate with compromised systems inside the network. Also known as a CnC or C2 server.

cyber threat intelligence: Knowledge about adversaries and their motivations, intentions, and methods that is collected, analyzed, and disseminated in ways that help security and business staff at all levels protect the critical assets of the enterprise.

distributed denial of service (DDoS) attack: A cyberattack intended to disable a targeted network or host by flooding it with requests from multiple computers.

hacktivist: A threat actor who uses cyberattacks to express political or ideological beliefs or to damage opponents.

honeypot: an Internet-connected computer that simulates the activities of servers or users in order to collect malware files and emails used in attack campaigns.

incident response (IR) team: The team responsible for investigating and analyzing data breaches and other cyberattacks. Also known as a computer incident response team (CIRT) or a security incident response team (SIRT).

indicator of compromise (IOC): An artifact or event associated with attacks or data breaches.

mass attack: An attack launched at a large number of potential victims rather than at a specific target.

network operations center (NOC): A facility for monitoring and controlling computer and telecommunications networks.

personally identifiable information (PII): Information that can be used to identify or represent individuals, including names, addresses, and financial and medical records.

pivot: (*verb*) To investigate a potential attack by starting with an initial indicator of compromise and finding related indicators and events.

sandboxing: Running an unknown file in an isolated virtual execution environment in order to detect malicious behaviors. A form of dynamic analysis.

security information and event management (SIEM): A system or application that collects and correlates security alerts and events.

security operations center (SOC): A facility for monitoring security alerts and events, initiating investigations, and remediating damage.

signature: A unique identifier of a file or other artifact potentially associated with an attack.

software development kit (SDK): A set of software development libraries and tools that facilitate the integration of an application with other programs.

spear phishing: Phishing campaigns directed at selected individuals within a targeted organization.

tactics, techniques and procedures (TTPs): Patterns of activities and methods associated with specific threat actors or groups of threat actors.

tradecraft: Operational techniques used in intelligence to obtain information from adversaries without detection.